Architecture Art Design Fashion Photography Theory and Things

With thanks for their
encouragement

Akira & Keiko Azumi
Yoshio & Yuko Nakajima

Anna Albright
Luke Hughes
Hiroshi Kaneko
Peta Levi
Dario & Romano Marcato
Hans Maier-Aichen
Chiaki Mikawa & abode
Issey Miyake
Yoichi Nakamuta
Eric Parry
TOA Co. Ltd.
Alexander Payne
Duncan McCorquodale
Maria Beddoes & Paul Khera

Black Dog
Publishing Limited
PO Box 3082
London NW1 UK

Available
Michael Young

Forthcoming
Tom Heatherwick
jam
Marc Newson

For further information
concerning other titles in the
Serial Books Design and
Serial Books Architecture &
Urbanism series,
please contact Black Dog
Publishing Limited in writing.

© 1999 Black Dog
Publishing Ltd &
Shin & Tomoko Azumi

All rights reserved.
No part of this publication
may be reproduced, stored
in a retrieval system, or
transmitted, in any form or
by any means, electronic,
mechanical, photocopying,
recording, or otherwise,
without the prior permission
of the publisher.

Every effort has been made
to trace all copyright holders,
but if any have been
inadvertently overlooked the
publisher will be pleased to
make the necessary
arrangements at the first
opportunity.

British Library cataloguing-
in-publication data.
A catalogue record is
available from the
British Library.

Serial Books Design was
conceived, edited and
produced by Duncan
McCorquodale

Designed by
Maria Beddoes &
Paul Khera
assisted by
Owen Peyton-Jones

Printed in the
European Union

ISBN 1 901033 02 3

photographs:
Julian Hawkins
Upright Salt Shaker
Funnel Top Pepper Mill
Dish Bottom Pepper Mill
Upright Salt Shaker (bone china)
Bench = Bed
Stool = Shelf
Armchair = Table
Wire Frame Stool = Shelf
Wire Frame Chair and Stool
Azumi portrait

Robert Walker
Snowman Salt & Pepper Shaker

Studio Synthesis
Overture / Screen = Cabinet
Coat hanger+Clock+Shelf

Thomas Dobbie
Table = Chest

Ed Reeve
Wire Frame Reversible Bench

Selected Exhibitions

1999
Salone Satellite Milano
Lost & Found British Council
Touring Exhibition
Stealing Beauty, ICA, London

1998
100% Design, London
Sinn & Form, International
Design Zentrum, Berlin
Powerhouse:UK, London
Decorative Art Today,
Bonhams, London

1997
Furniture Futures by Design
Week, Selfridges, London
100% Design, London
Flexible Furniture,
Crafts Council, London
From Prototype to Production,
Ruthin Craft Centre, Wales
*Contemporary Decorative
Arts,* Sotheby's, London

1996
Objects of Our Time,
Crafts Council, London
100% Design, London
Design Resolutions, Royal
Festival Hall, London
Decorative Art Today,
Bonhams, London
*Design of the times: 100
years of the Royal College of
Art*, RCA, London

Work in public collections

Victoria & Albert Museum,
London
Geffrye Museum, London
Crafts Council, London
Stedelijk Museum,
Amsterdam

Bibliography

1965
Shin Azumi
Born in Kobe Japan

1989
BA in Product Design at Kyoto
City University of Art

1989 Grand Prize of Seki
Cutlery Design Competition,
Japan

1989–92
Works for NEC Design Centre
Co. Ltd. Personal Computer
Department, Japan

1991, 1992 Good Design
Award by JIDPO, Japan

1994
MA in Industrial Design at the
Royal College of Art, London

1966
Tomoko Azumi
Born in Hiroshima Japan

1989
BA in Environmental Design
at Kyoto City University of Art,
Japan

1989
Misawa Student Housing
Design Award, Japan

1989–90
Works for Kazuhiro Ishii
Architect and Associates,
Japan

1990–92
Works for Toda Construction
Corporation, Design Room,
Japan

1993
ABSA/Arthur Andersen
Trophy Design Award, UK

1995
MA in Furniture Design at the
Royal College of Art, London

FX/HNB Furniture Award at
the New Designers Exhibition,
UK

1995
Azumi formed

1997
Chair, Student residence
Pembroke College,
Cambridge University
Cambridge

Collaboration with Eric Parry
Architects, and furniture
manufacturer, Luke Hughes &
Company

1998
Display Table / Bench,
High Stools
Joanna's Tent London,

Meeting Table, Computer
Desks, Shelving System
Commission East,
Cambridge

Stacking Bench / Shelf and
Stacking chairs
Kettle's Yard Gallery
Cambridge

2000
Cafe Furniture
Display Cabinet
Folkestone Library &
Museum Cafe
Folkestone
collaboration with architects,
Adjaye & Russell, and artist
Chris Ofili

Spa Furniture, Mandarin
Oriental Hotel
Commission to develop a set
of site specific furniture for
Spa in the Mandarin Oriental
Hotel, London. Furniture
includes, chairs, day beds, and
seating. Collaboration with
Eric Parry Architects, and
furniture manufacturer, Luke
Hughes & Company, London

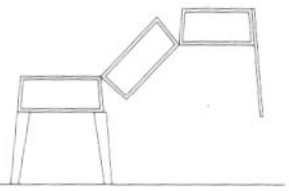

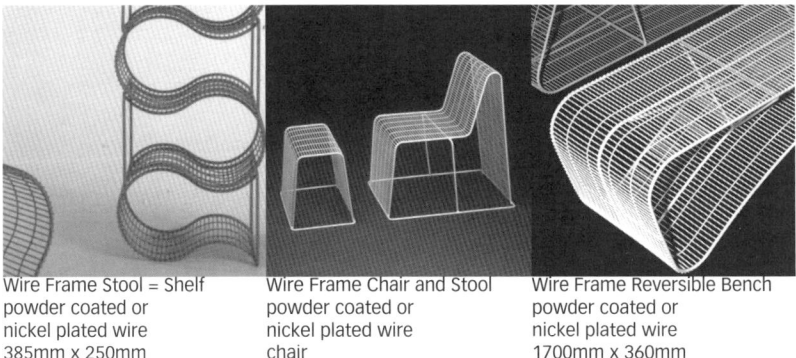

Wire Frame Stool = Shelf
powder coated or
nickel plated wire
385mm x 250mm
x 380mm high
Azumi
1998

Wire Frame Chair and Stool
powder coated or
nickel plated wire
chair
550mm x 700mm
x 730mm high
stool
500mm x 320mm
x 380mm high
Azumi
1998

Wire Frame Reversible Bench
powder coated or
nickel plated wire
1700mm x 360mm
x 400mm high
Azumi
1998

Table = Chest
beech veneered plywood
solid beech and
metal fixture
table
900mm x 400mm
x 400mm high
chest
400mm x 400mm
x 720mm high
Azumi 1st edition
abode, Japan, 2nd edition 1998
Tomoko Azumi
1995

Folding Stool
350mm x 150mm
x 560mm high
limited batch production 1996
Tomoko Azumi
1993

Stool = Shelf
solid maple and steel pipe
350mm x 320mm
x 430mm high
limited batch production
Azumi
1997

Armchair = Table
maple veneered MDF,
solid maple and steel pipe
armchair
580mm x 620mm
x 830mm high
table
580mm x 900mm
x 700mm high
limited batch production
Azumi
1997

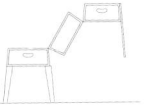

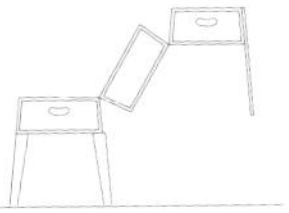

Moon House
computer graphics
Tomoko Azumi
1989

Paper Light
handmade recycled paper,
cables, low voltage halogen
system, transformer and
stainless steel wire
450mm x 300mm
x 4000mm high
one-off
Tomoko Azumi
1993

Overture / Screen = Cabinet
maple, powder coated steel
and cotton
520mm x 520mm
x 1850mm high
Azumi 1st edition
Lapalma, Italy 2nd edition 1997
Tomoko Azumi
1995

Bench = Bed
MDF, beech and metal hinge
bench
1900mm x 485mm
x 540mm high
bed
1900mm x 800mm
x 270mm high
limited batch production
until 1998
Azumi
1996

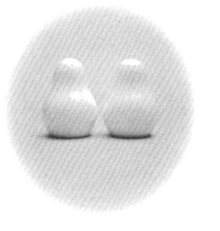

Funnel Top Pepper Mill
beech, American walnut and
pepper mill mechanisms
76mm Ø x 110mm high
Shin Azumi
1995

Dish Bottom Pepper Mill
beech, American walnut and
pepper mill mechanisms
flat type
100mm Ø x 50mm
sphere type
100mm Ø x 60mm
Shin Azumi
1995

Upright Salt Shaker
(bone china version)
bone china and cork
50mm x 110mm high
Azumi
Shin Azumi
1996

Snowman Salt & Pepper Shaker
58mm Ø 67mm high
Authentics, Germany
1999
Shin Azumi
1998

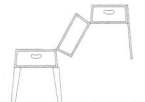

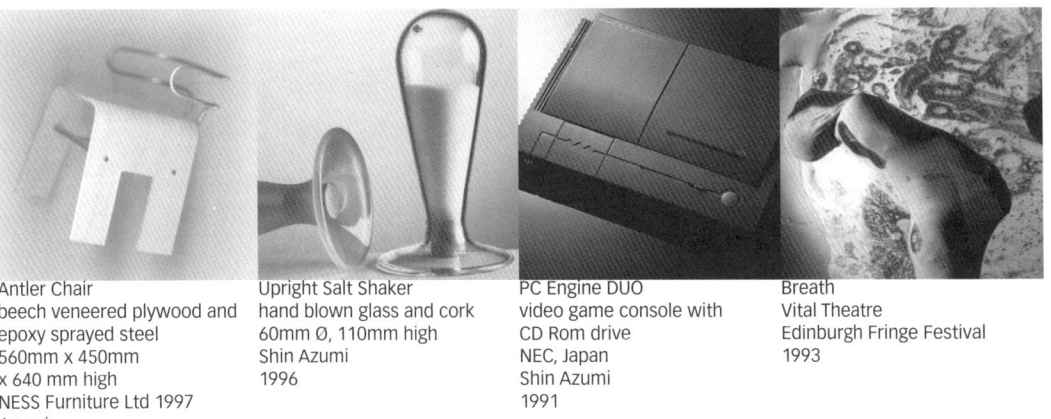

Antler Chair
beech veneered plywood and
epoxy sprayed steel
560mm x 450mm
x 640 mm high
NESS Furniture Ltd 1997
Azumi
1996

Upright Salt Shaker
hand blown glass and cork
60mm Ø, 110mm high
Shin Azumi
1996

PC Engine DUO
video game console with
CD Rom drive
NEC, Japan
Shin Azumi
1991

Breath
Vital Theatre
Edinburgh Fringe Festival
1993

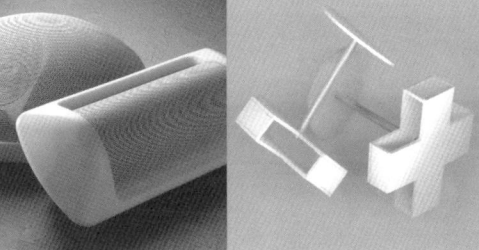

Radio = Window
Solar Powered Radio
solar cell, plastic, lycra, metal
tape and suction pad
120mm x 60mm x 160mm high
maximum length
of the aerial 900mm
prototype
Shin Azumi
1994

Coat hanger+Clock+Shelf
maple veneered plywood,
clock mechanism
Lapalma, Italy
Azumi
1998

H-1 Speaker (Cylinder Type)
ABS and steel
312mm x 126mm x 90mm high
TOA Co. Ltd Japan
Shin Azumi
1999

H-2 Speaker (Dome Type)
ABS and steel
288mm x 111mm high
TOA Co. Ltd Japan
Shin Azumi
1999

Cross Table
steel, plywood and
plastic laminate
600mm x 600mm
x 700mm high
Sumitomo Bakelite Co. Ltd /
Trunk, Japan
Azumi
1999

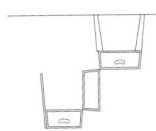

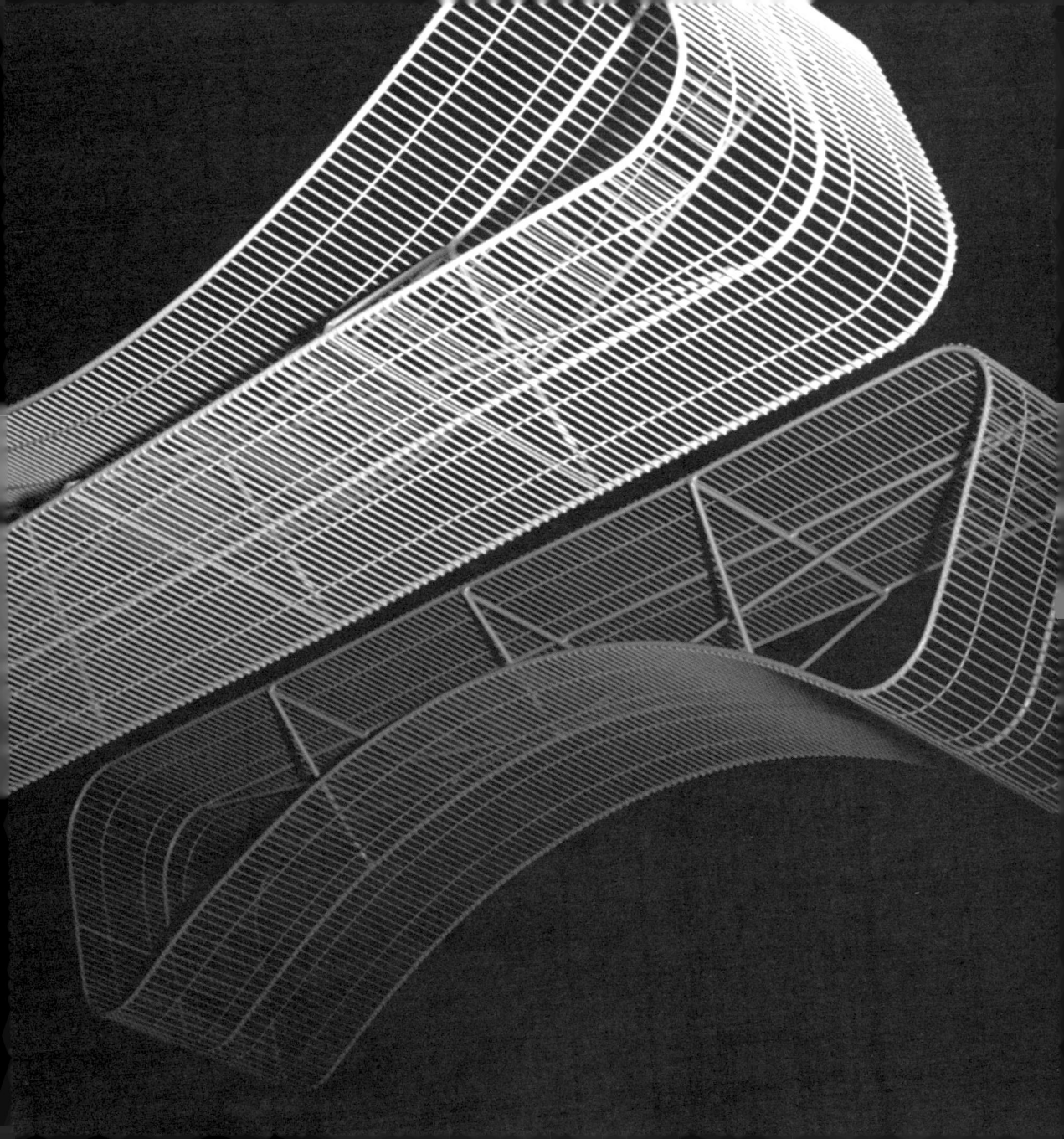

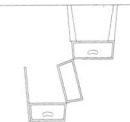

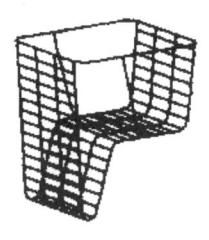
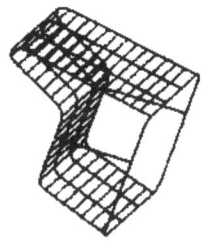
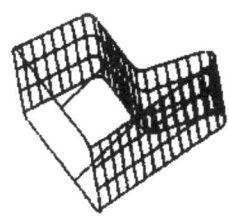

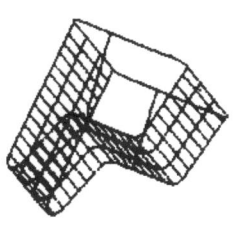
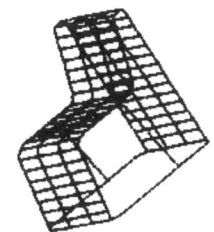
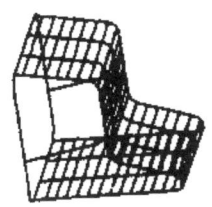

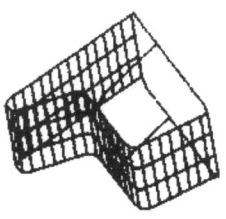

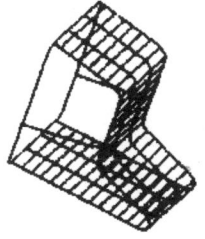

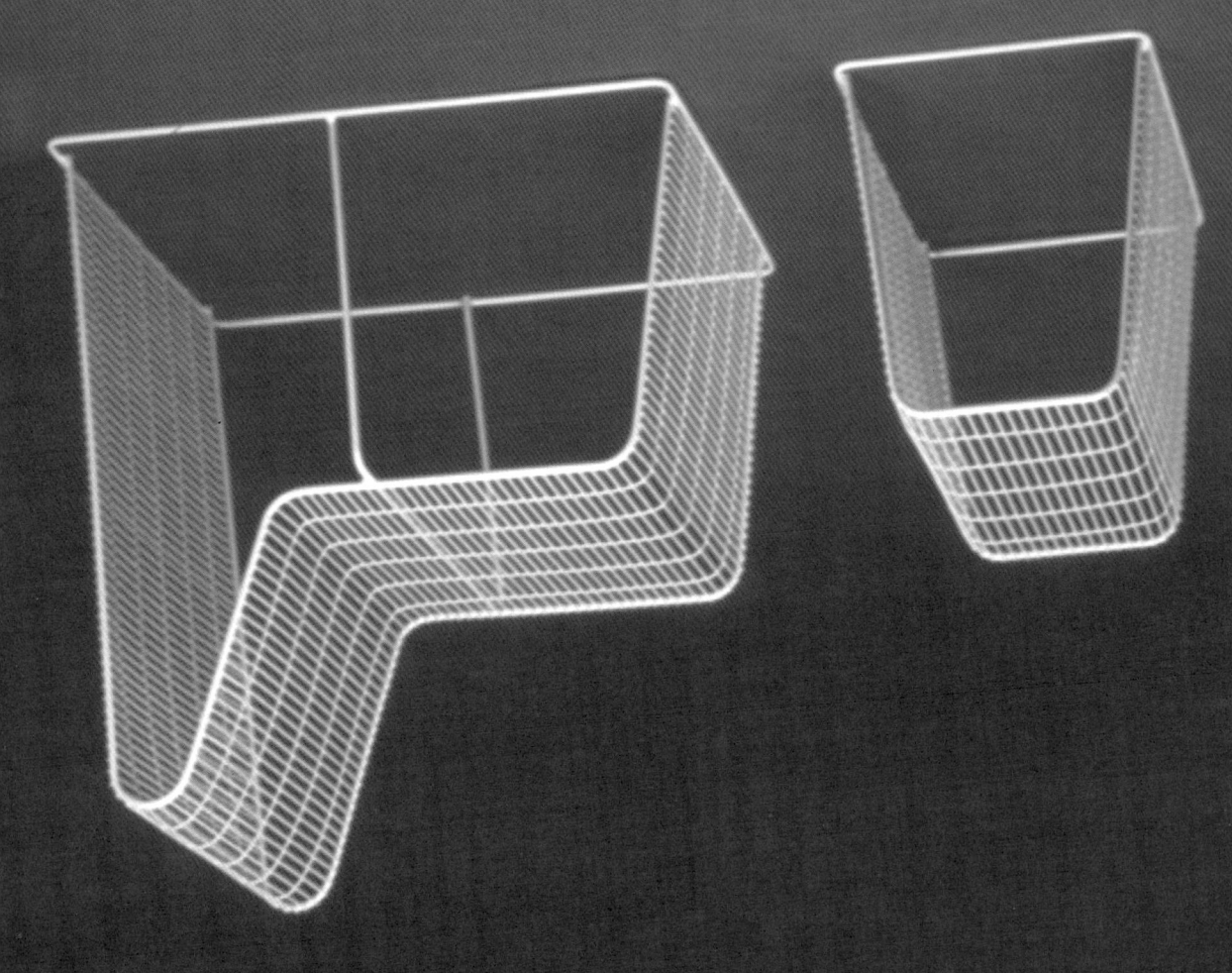

In the 'Wire Frame' series we looked at inexpensive materials and production processes. We were looking at supermarket trolleys and baskets, which we thought would be reasonably inexpensive to produce and could be made with available technology. We thought the translucent quality and 'visual lightness' of this material could satisfy functional requirements without occupying a lot of space. You can always see the edge of a room through the furniture, which creates less of an impact....

In our imagination, the material overlapped with the image of a primitive wire frame 3D computer drawing — a 3D object in a 2D world. It has no gravity or physical existence, but it has an impact.

In the 50s, the Eames', Bertoia and Panton, amongst other modern designers, had already used these manufacturing processes in their work. These designs were beautiful but used a very expensive, labour intensive technology. What we have tried to do is make something using simple, prefabricated methods.

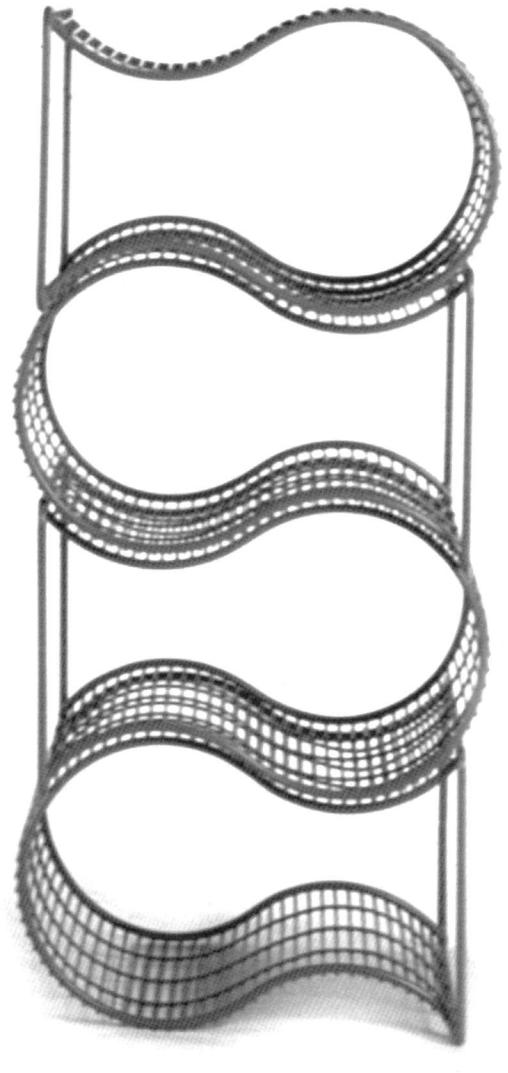

It was during the process of developing this new furniture series that we found 'transforming movement' to be a 'fun' element. The dual purpose to this furniture is another key.

So, the special thing about this furniture is, in fact, that a 'chest' becomes a 'table'. They have a completely different character in the end. At this time, we didn't know why, but people had not developed this area of design very much.

We know why now! — the marketing reasons...

It's very difficult to sell a piece of furniture with two names, such as *Armchair = Table.* The concept behind the design does not go through to some retailers very well... many of them hesitate to take this kind of work.

We weren't aware of that problem. We were just doing what we wanted, but after we approached manufacturers we started to realise why people avoided these areas.

But this, then, became another reason for us to expand in this area — a kind of untouchable area in the furniture industry.

The British audience did not take the same approach to our work as the furniture industry. They enjoy our innovative ideas... the 'category' was not really a problem for them.

At the RCA the tutors told us to "push the boundaries of the furniture". My response was a bit abstract, but I took the words in this way — "Perhaps I can remove the category of furniture or its limited definition, such as meaning 'bookshelf' or 'partition screens'. "how can I push the boundaries of the furniture literally. I could combine two or more items together.

When I made the very first model of *Table= Chest* in paper it looked awful. But it's movement started to suggest something interesting, a different quality, something that could be called 'fun' — a balance between function and transformation. It was only after I finished this piece that we seriously started to analyse the balance.

I used to be very much against it if someone said my design looked Japanese. But I've started to accept this as it is clear that my work reflects something of my background and culture. At some point, I thought — OK, I will accept this. I've even thought that it might be used in some way...

We have this discussion of Japaneseness when we undertake a project. But we also understand that Eastern ideas and styling bring something 'new' which might be accepted... there is a certain market for this, but we have to be very careful because we want to show our furniture in Japan too. Ideally, we would like to find and create some universal shape or language, which would be accepted all over the world. We think the concept of 'dual purpose' is already more universal than Japaneseness.

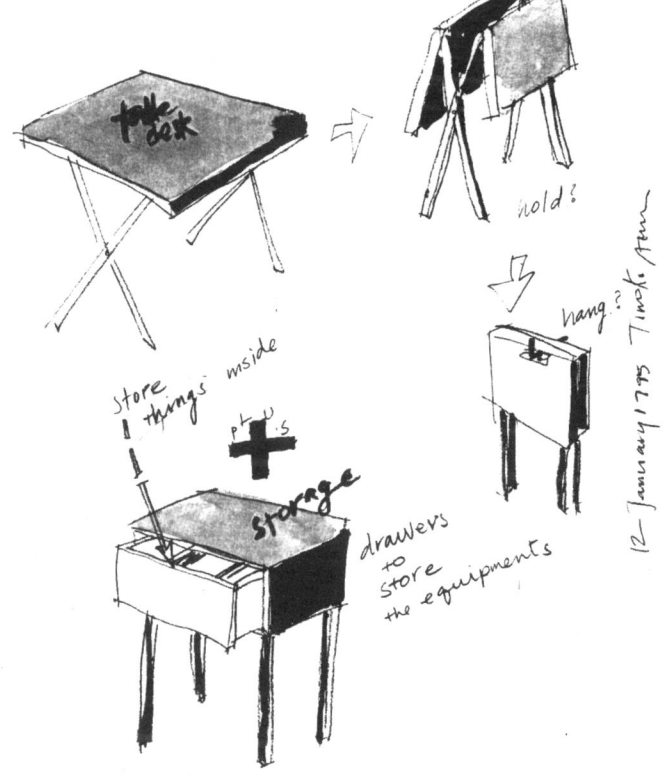

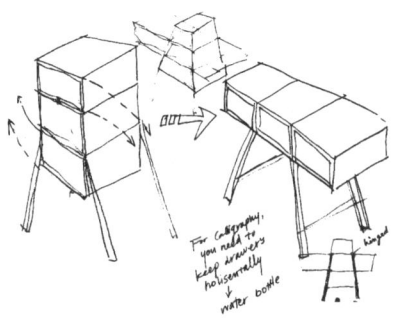

What had we done in designing this piece? Then we thought 'dual purpose' — a solution to problems both we and our audience share in modern life. 'Japaneseness', too, might be another factor, but really we carefully avoid obvious Japaneseness in these pieces.

Table = Chest
was received very
well, better than
we expected.

At the time I thought the design was original and
quite an interesting transformation... but we didn't
think it would be received so well. So, we started to
ask ourselves why?

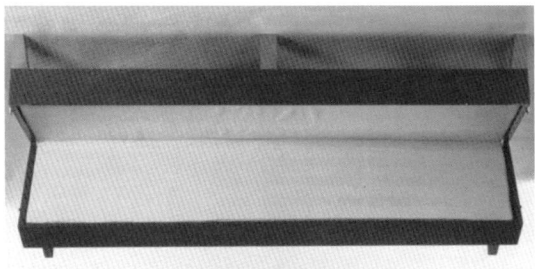

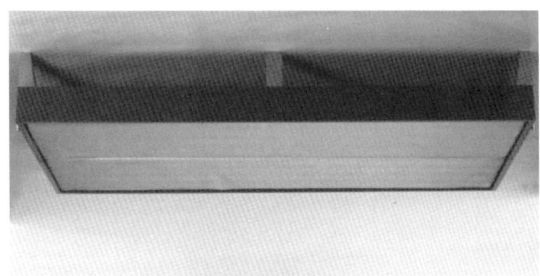

I had to work things out seriously otherwise the room would become very much smaller. If I bought a bigger TV, then I would have to get rid of something in order to squeeze the TV in. That was my basic experience of how to put things together, how to manage. Sometimes I tried to move elements of the room around. I would then think "Ooh, that's much better, but why?". I wanted to know the secret, which was another reason for me keeping this record. So, I can see from this notebook that I first introduced a chair to my room in about 1991. It was a light fold-away chair that I used mainly when I needed to reach up high for storage. I mostly lived on the floor at this time, which is the best solution for mobility. Chairs become important to our lives much later, after we moved to London... for me they are part of a Western way of life.

Obviously, you can see a connection between our experiences and this bench, which turns into a bed. We designed this because we had many occasions where we needed this type of furniture... need is the underlying reason, but routines are also important. For example, if you move this *Overture Screen* a TV appears from behind it. And if you move chairs towards the TV and sit in front of it, the TV becomes a small cinema — animation and motion...

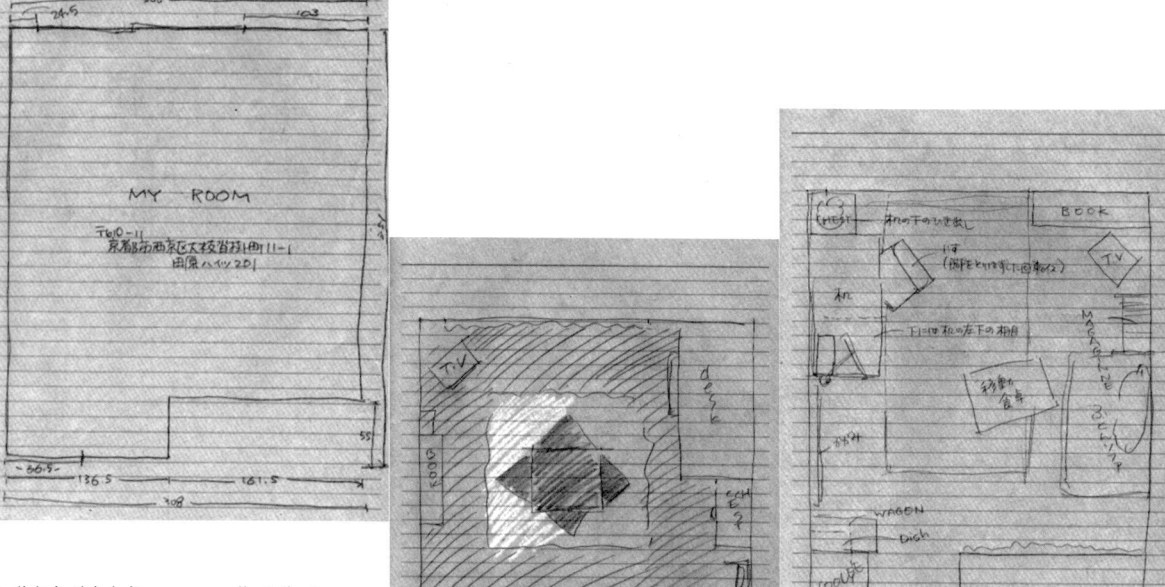

I didn't think it was small at that time. I started these notes, a memo really, to remember what I had done with the space. It was, in a sense, almost a mental space, because all this took place within nine square meters. When I moved to a smaller room in Tokyo I had to store things in more of a systematic way. I designed a low, built-in storage unit and hid things behind a curtain. As I had started to do my own personal design projects in these rooms,

flexible and fold away furniture became a natural and inevitable design choice for me.

INTERIOR NOTE

My experience of living in Japan connects to our current works. 'Reality' there for me was, for example, living in small rooms from the time I left high school and started living in Kyoto.

This is the notebook of my room.

which I started in 1985. The first room I had in Kyoto was less than ten metres square. That was a normal size for student residences. I actually lived in a smaller room when I was working in Tokyo. The rooms were just for sleeping really, but I wanted to make the space comfortable. Just beside the entrance to a typical room is a bath unit of industrial plastic, moulded with a small bath tub, toilet and washing basin in-between. There is a small kitchen too. I was very pleased that I had my own space...

While at The RCA I designed a small mobile shelter — the *Overture Screen = Cabinet*. It involved a lot of compromise, but, somehow, I wanted to create a shelter like that of an insect's shelter before it turns into a butterfly — a functional object. I like the idea of being sheltered in a translucent skin, seeing the changes in the environment through light. This is kind of practical realisation of my 'translucent skin' idea. It was originally called *Screen=Cabinet*. I designed this for my degree show at the Royal College of Art.

There was another influence on it's design too, this was the performance *Breath* I did with Shin at the Edinburgh Festival. For this performance we used a big cotton cloth as a screen on which to project images. I was acting as a stage hand in the performance and always sat behind this screen. Here I saw strange and beautifully distorted moving images, in shadow, projected against the screen. I wanted to include something of this kind of movement in a piece of furniture.

I was fascinated by Rudofsky's *Architecture Without Architects*. I thought I would use solutions to problems of climate — a regionally cultivated kind of wisdom, such as the tent architecture of native North Americans — the idea that you can carry shelter with you while on the move.

I was also interested in the work of Buckminster Fuller. Yes... quite idealistic, not, in my mind, very practical at all! I couldn't really believe Bucky's philosophy, it's shades of socialism. I don't believe in that part of his thinking — that humans could create a 'climate' within a dome... I thought this to be a crazy idea! I was fascinated, though, by his ideas of wanting to understand more about the earth and his progressive approach in dealing with the future.

The idea of 'mobility' also somehow touched me emotionally at this time.

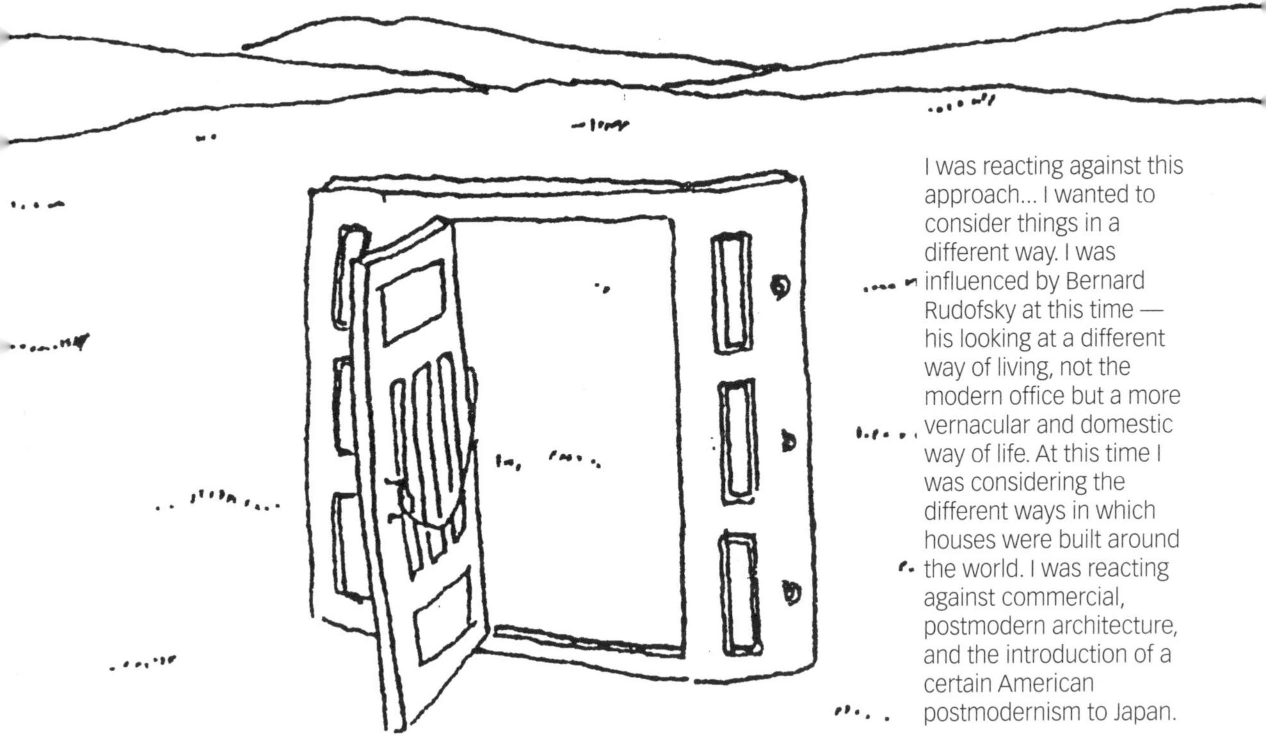

I was reacting against this approach... I wanted to consider things in a different way. I was influenced by Bernard Rudofsky at this time — his looking at a different way of living, not the modern office but a more vernacular and domestic way of life. At this time I was considering the different ways in which houses were built around the world. I was reacting against commercial, postmodern architecture, and the introduction of a certain American postmodernism to Japan.

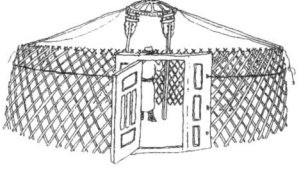

At that time your hero was the architect Toyo Ito...

Yes, in the late 70s... yes, he started to do translucent architecture and some interesting furniture, in 1985, called *Pao for Nomadic Tokyo Girls*, which was a kind of shock to me.

But I was generally suspicious of architecture's methods, I wasn't convinced by the way architects designed buildings. It was theoretically advanced, but I couldn't agree with the way the client's needs weren't reflected in his architecture.

In Japan, because the sun is much higher, its much stronger... as a result we have always had the need for some kind of shade, such as Shoji screen.

But even if you are deep within a house you can still see the light coming through to its interior. I remember well this sort of scene from my childhood.

When I first saw work from the Bauhaus I reacted in different way to Shin. I was interested in Moholy-Nagy.

These see through type images layers and strips of light and shadow. I've always been fascinated by this type of thing...

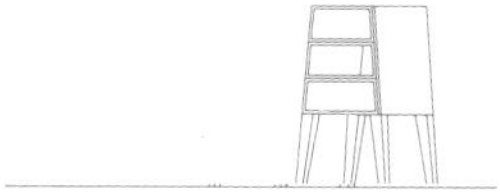

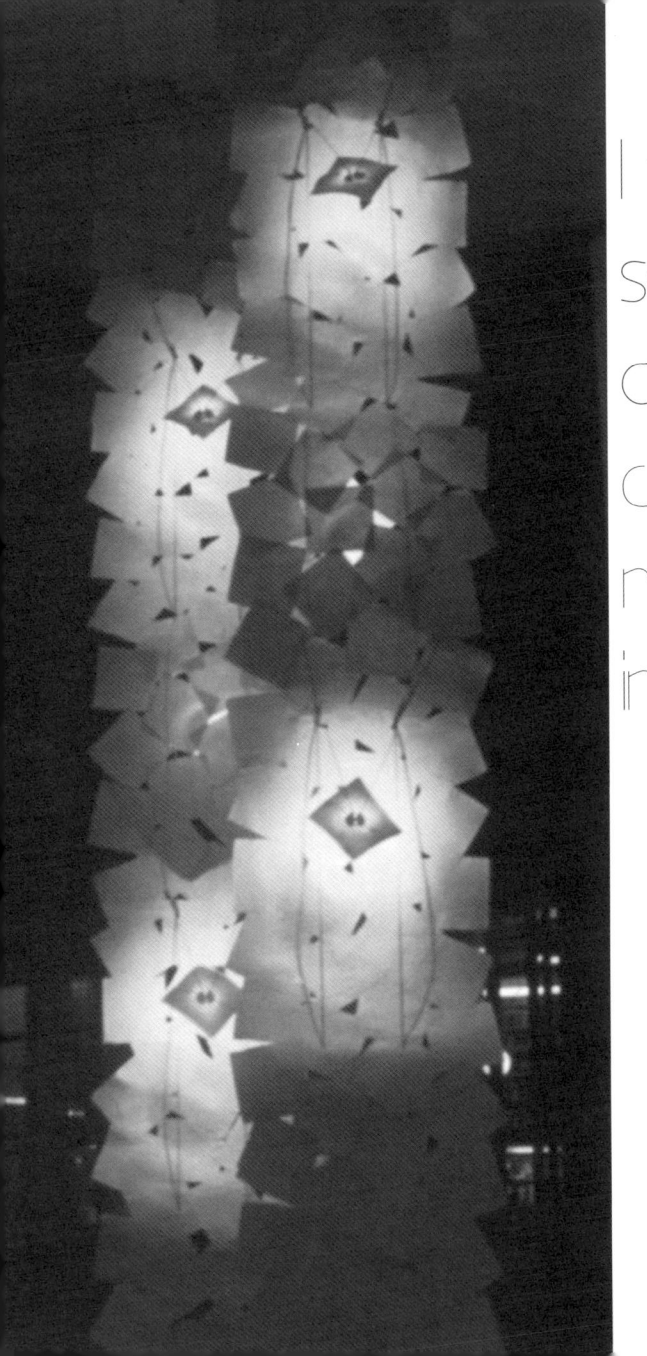

I wanted, somehow, to combine the tactile quality of paper, movement and light into furniture.

This is a computer model of a house I designed in the final year of my study ay Kyoto, called the *Moon House*. You could follow the movement of the moon from east to west using the windows or the reflections created by the small pond in front of the house.

This was very much a postmodernist moment, a period during which people were doing really crazy things....

When I came to London with Shin he already had a place at the Royal College of Art. I applied to study architecture there, but I couldn't get a place. I then, took a one year course at Goldsmith's College. This was a prepatory course, comprising English, Art History and Studio. It was at Goldsmith's that I was introduced to tutorials, which was a very new experience to me. I was encouraged and allowed to do anything I wanted. I used paper a lot at this time. I like paper as material — its smell, fragility, translucency and tactile quality. I wanted to somehow use paper in space. So, I made small objects, which looked like a fan. I then connected a lot of these together and hung them in space....

Then, from 1993, I studied design at The Royal College of Art. In the first year of my studies at The RCA I designed seating and some folding furniture. I also designed a screen — something you can use as screen and for storage purposes. I was, as well, very interested in designing blinds, and, again, I started to make paper. I used recycled paper and made a sort of blind, the angles of which you can change so as to control and adjust the amount of light coming through it.

Our first real collaboration was a group work done during the first year of my university studies in art at Kyoto.

The idea behind this collaboration was the making of a kinetic object, an object which moved with the wind. This work was very much influenced by the sculptor Susumu Shingu.

At Renzo Piano's Kansai Airport Shingu installed a wind sculpture in one of the big foyer spaces, just below the ceiling of the space... Piano's design incorporated a big extraction fan in this ceiling, so there is always a 'wind' in this area... and the sculpture is always moving. It's beautiful.

... floating, rotating, unpredictable movements, but never touching the ceiling. It's very controlled...

I was very much concerned with architecture at the beginning of my study. I took a course in Environmental Design at Kyoto City University of Art. The course consists of furniture, interior design and architecture. I was interested in designing a house, a domestic environment. I had been interested in this for years, since I was a child, in fact. I had collected a lot of small plans of houses, these were like apartments or flats.... And my father was an influence in some way too. He is a structural engineer. I used to watch buildings being constructed on site with him and I really liked what was going on. But when I came to study architecture I concentrated more on domestic concerns, on houses.

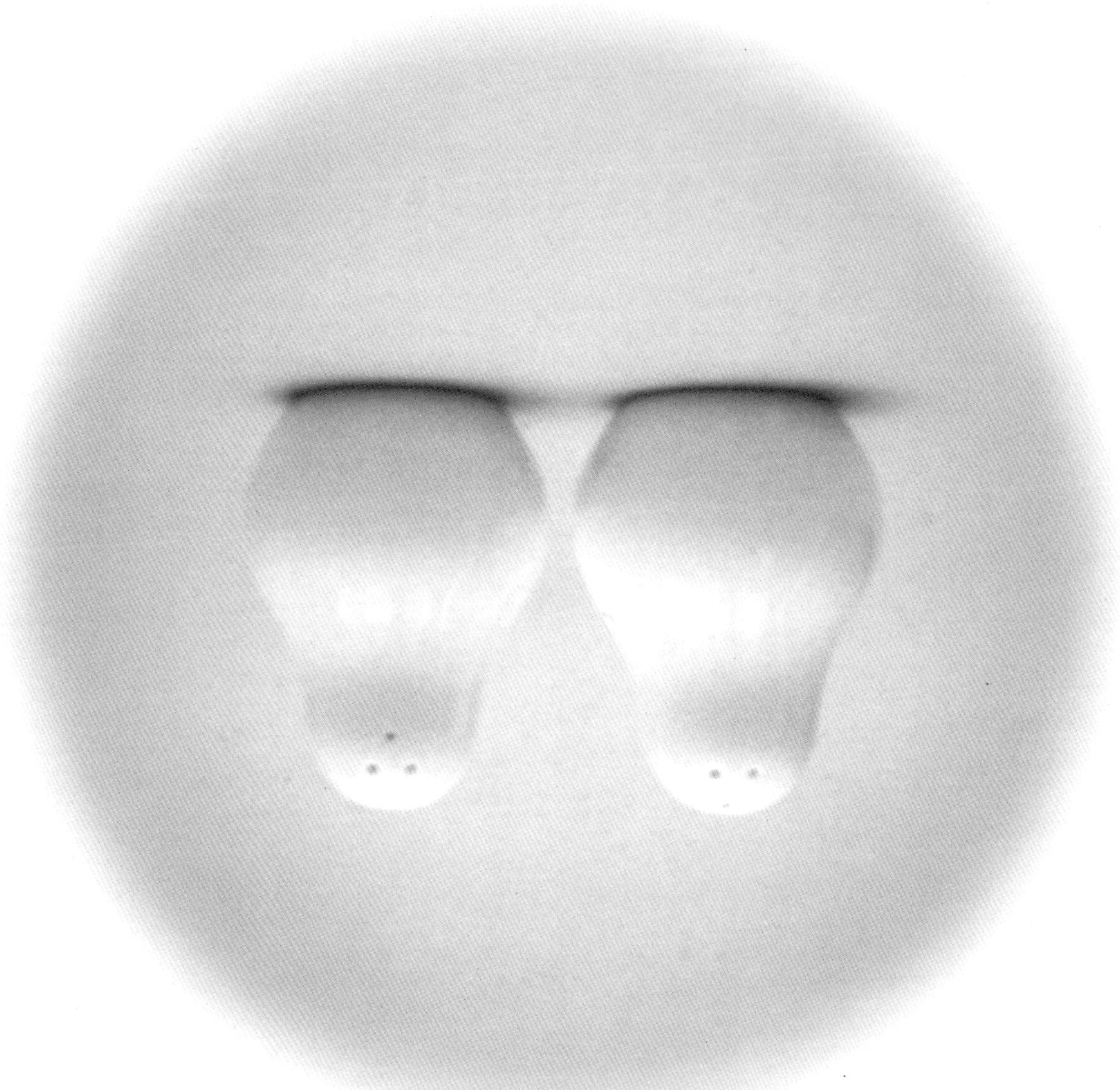

After I had changed the material from being transparent to opaque, I noticed that the combination of the shape and the hole started to create a sort of human expression. According to psychologists, if we see two dots in a horizontal line we start reading a facial expression. The next step, then, saw me starting to play with the expression resulting from the combination of the shape and dots.

I had created the Snowman Salt & Pepper Shaker.

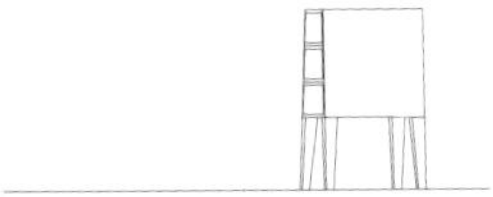

The tableware series shows a process of development in my thought. I started with *Funnel Top Pepper Mill,* which is about functionality. I then made this 'dish bottom' pepper mill — the result of recognising another problem with conventional pepper mills. These usually leave grains of pepper behind because the grinding mechanism is on the bottom. I simply covered this mechanism with a dish. After I designed this version of the pepper mill, I noticed that I had changed the manner of using this piece of tableware. Using this mill I am no longer grinding pepper in a conventional way. I thought I had come upon a very interesting discovery. I then tried to improve on this further in a third object, the *Upright Salt Shaker.* I tried to create a better way of shaking the *Shaker.* But the first design was too expensive as it was made of hand-blown glass. So, I made a second version, *Upright Salt Shaker,* from bone china.

When I started designing and making batch productions of wooden tableware, I was simply looking at, and thinking about, details of my daily life. Everyone feels these, but they aren't really present. So, I thought, how to improve daily life in a creative way?

This 'funnel-top' pepper mill, for example, allows you to easily refill it with pepper corns. Such a design simply makes daily life better. Everybody has this problem refilling their pepper mill. It's an irritating moment in the day. Of course, it's just a small detail and people get used to the problem. I think this is funny.... People are sharing this irritating experience and yet it is not pointed out.

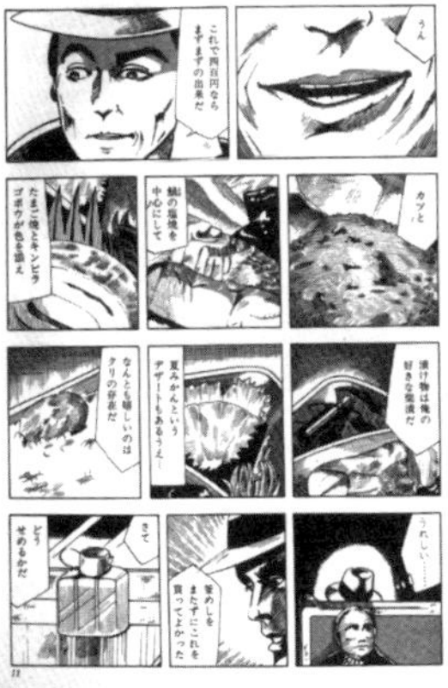

Izumi is very interested in intimate, everyday behaviour, behaviour which people are not usually aware of as it is too familiar. His early book *Groovy Sukiyaki* is all about food, ways of eating and the psychology around them. *Night Train*, another of Izumi's works, is a comic about eating a bento box (a lunch box) on an overnight train. The character in *Night Train* is trying to find harmony in what he is eating. He is analysing and planning each combination, the order in which he eats his food. Towards the end of the story his plans start to collapse, finishing in tragedy. *Night Train* is really a tragic comedy.

This kind of experience is familiar to me, but it has never been pointed out with such preciseness.

かっこいいスキヤキ

泉昌之

The movie Tampopo is
another of my
favourites. It's about the
noodle, food, manners,
behaviour and an
enthusiasm around
these.... The film is very
much influenced by the
manga comic author
Masayuki Izumi.

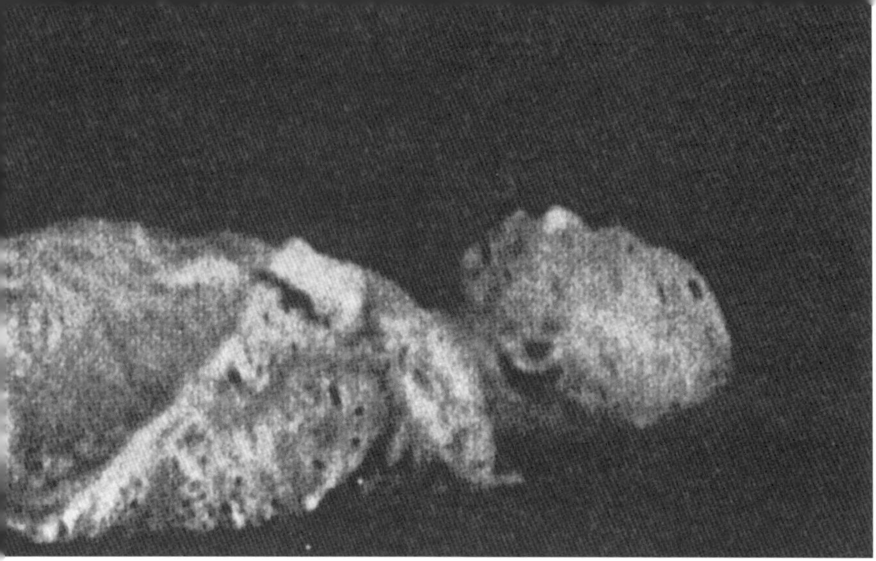

At the same time we started looking at performance theatre we were fascinated by Butoh — a new wave of Japanese dance that started in the 60s. Works by Sankai-Juku, in particular, expressed, for us, the pure existence of the human body. Sankai-Juku, provide perhaps the best example of work that started from Japanese origins but was sublimated, becoming something more than Japanese.

We have also been influenced by the theatre of the Canadian director Robert Lepage. The quality and the quantity of ideas in his work is amazing. In 1992 he directed *A Mid-Summer Night's Dream* at The National Theatre. For this staging he chose a female contortionist for the part of Puck. The contortionist's unusually flexible body was very expressive and was used to effect in successfully luring the audience into the world of the fairy.

Lepage also introduced a great many ideas into the design and the effects on the stage. He used water, for example, reflections in water, as a beautiful way of working with light.

His stages were designed in a very simple but clever way — not gorgeously, like musical theatre, but changing quickly and dramatically.

Tomoko

This is the same with our furniture. We want to express pure thought. We want to hide cultural 'things' behind an idea. And we don't want to be associated with an obvious 'Japaneseness'. We want to concentrate on the functional aspects of design, not its decorative features.

Yes, this is our attitude. We are trying to concentrate on the most important message or idea in our designs, getting rid of the others....

We want to discover universals, even though these might be based on a very localised, everyday life

We are looking to basic 'things' that we might be able to share internationally.

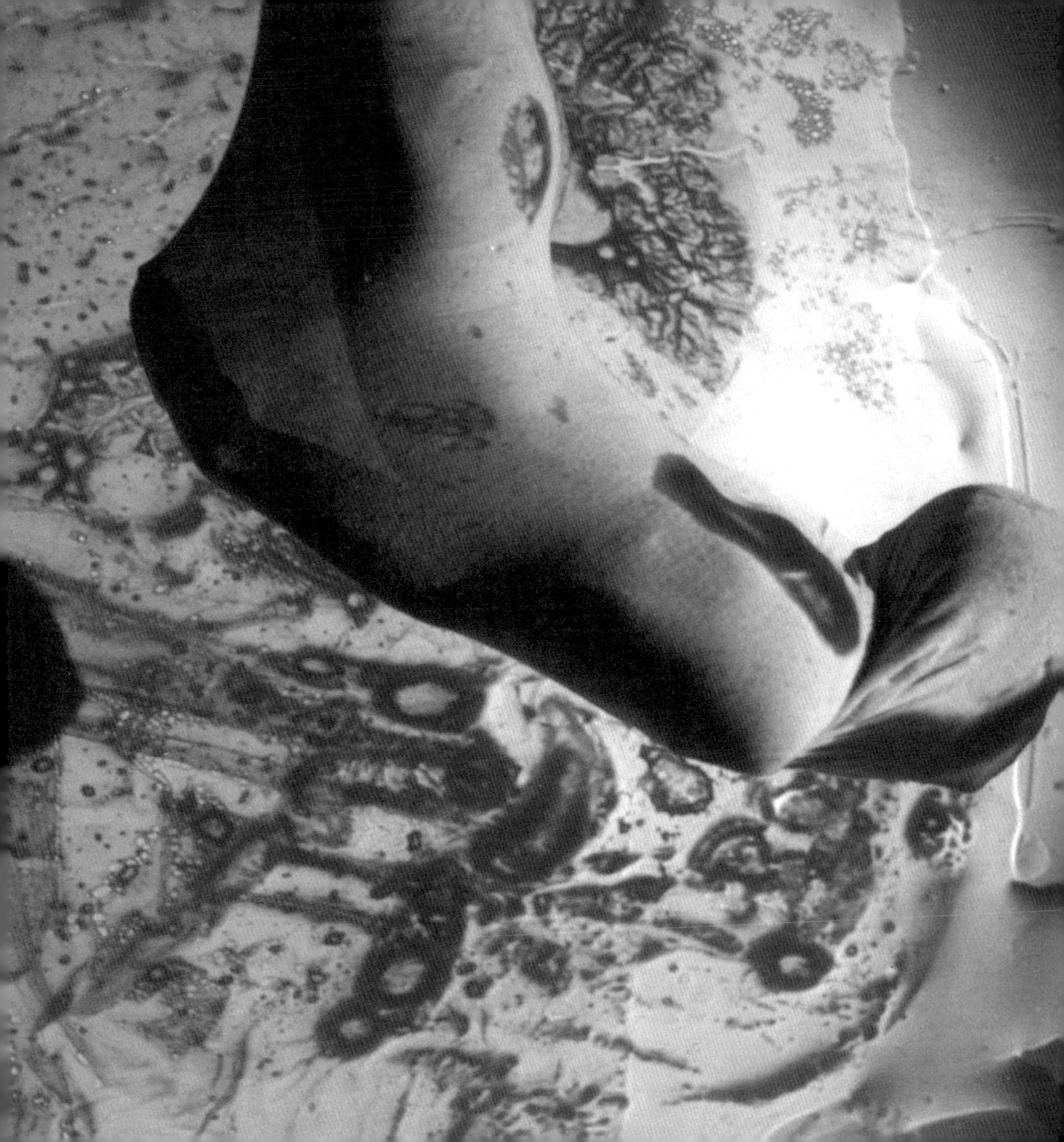

After finishing my first year at the Royal College of Art Tomoko and I formed a small performing group, along with students from Laban Centre and Goldsmith's College. We created a theatre piece called *Breath,* which we presented at the Edinburgh Fringe Festival in 1993.

The starting point of *Breath* emphasised the 'live' aspects of performing art — We tried to stage a performance that would present a purified, raw energy. We were inspired in this by primitive underwater animals, such as jelly fish and animalcules, the beautiful movement and fragility of these life forms.

The performers in *Breath* were covered by a flexible fabric tube. From within this tube they could sculpt and change their shape. Visually this resulted in there being no sense of front or back, top or bottom to the performers in the piece. We wanted to free them from the limitation of ordinary human expression, which is usually reflected in terms of gender, culture and age. In this we tried to convey a purified expression — the 'live' existence of the performers. Through our collaboration with the performers I became attentive to picking up and reflecting the movements of daily life in a work.

With *Breath* we attempted to hide the cultural aspects of movement, instead trying to convey a more purified expression of the body — like that of an animal. We knew this was going to be difficult, but we believed that it might be more universal, touching our audience with a different level of creativity if it were successful.

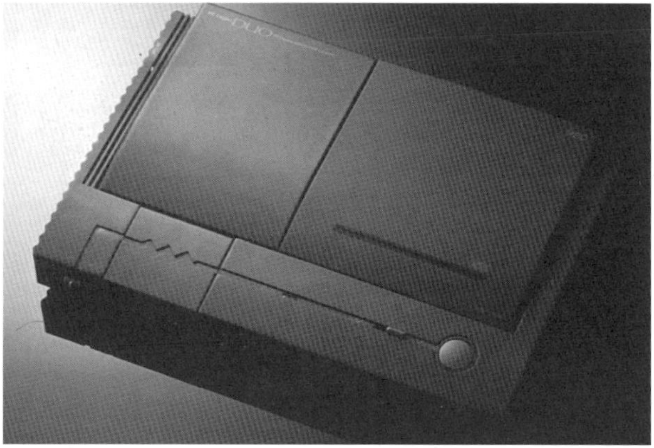

important because they are not user friendly products by nature. I was required to look at and analyse peoples' behaviour very carefully. If there was something wrong with a man-machine interface, I was required to prevent whatever it was that was going wrong from happening, to modify it. The video game console I designed is one example of an expression of function. It's 'graphic line' is like a pulse of energy. All of the console's functions are indicated on that line — on/off switch, software slot, LED, CD player lock/unlock button, etc.. So, the user can understand the idea of the object's use without literally being told about.

There is also another dimension to this which provides some suggestion of my interest in function and interaction within objects. A design I did while still in Japan, a video game console for NEC, borrows from the theory of "semantic product design"—an idea,I believe, coming out of Cranbrook University—which was a hot topic amongst Japanese industrial designers at the time. In short, semantic product design entailed an approach to the design of an object that expressed the function of the object directly. While I was at NEC, I designed computer related products, such as a video game machine, a personal word processor, a lap top computer, and so on. With these types of products a subtle suggestion of their function is very

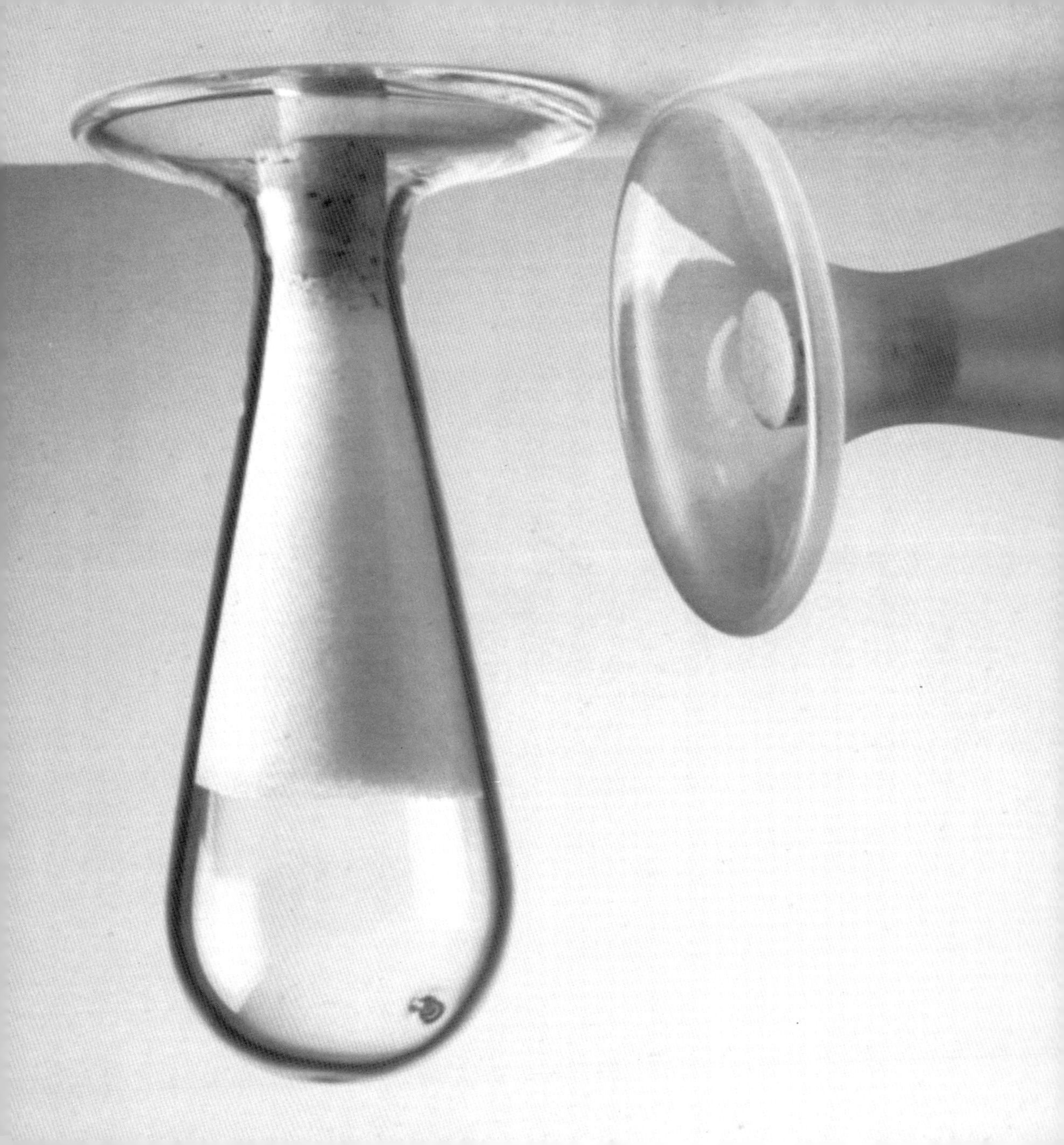

The easiest object to describe this quality of 'performance' is the

Upright Salt & Pepper Shakers.

There are holes piercing the head of each of these shakers. Usually this kind of hole is on the top. But I made holes slightly lower than in a usual shaker. This small detail only slightly changes the user's manner of using the object. With the *Upright Shaker* the salt or pepper comes out when it is slightly tilted. It isn't necessary to turn the shaker upside down or shake it aggressively. There is also a funnel on the bottom of each shaker, which makes them very easy to refill.

The *Upright Shakers* are about interaction and slightly changing peoples' behaviour , making it better, more refined.... This design is not just about function or the objects sculptural quality.

Imagine, if we start to produce this object in mass quantities and people start using them in a certain way — it's a kind of pop art, a performing art... exhibiting our furniture, people often simply enjoy watching how it changes. It is like a small puppet theatre.

My interests, at this time, started moving towards media that might bring about a quicker result, a more immediate response. I became interested in performance, in performance theatre. At this moment I was a university design student and, naturally, as a design student, I looked at Bauhaus theatre. Here I discovered the work of Oscar Schlemmer.

Schlemmer tried to simplify, reconstruct and shape the human body and the movement *around* the human body. This was a big discovery for me.

The Bauhaus theatre led me to become aware of design as a performing art. For example, the 'tableware' series is very much about our behaviour and interaction with an object. So, in a sense, if I improve the way an object is used I start to influence peoples behaviour in a particular way.

Changing peoples daily behaviour in a very subtle way could be described within a context of performing art.

Another lead into my work is through animation — the cartoon. There is a strong 'culture' of animation in Japan and I have been very influenced by this. When I was a teenager, I was given the opportunity to watch a film by Canadian animator Norman McLaren. He had created abstract, but very entertaining animations that were synchronised to sound... geometrical compositions and movement, synchronised movement with sound... They are fascinating. I suddenly became a great fan of McLaren's work. And I started to watch the work of 'experimental' animators as well. It didn't take long before I was making my own animations. However, after I had made several of these short films I began to feel frustrated with the process. I enjoyed creating funny moving pictures using film, but the process took too long. I wanted a quicker result.

After I started working with Tomoko, and our development of the 'flexible furniture' series, I remembered the movement in Norman McLaren's animations. When we designed Armchair=Table, the movement and transformation of the furniture completely overlapped with McLaren's *Canon* in my imagination.

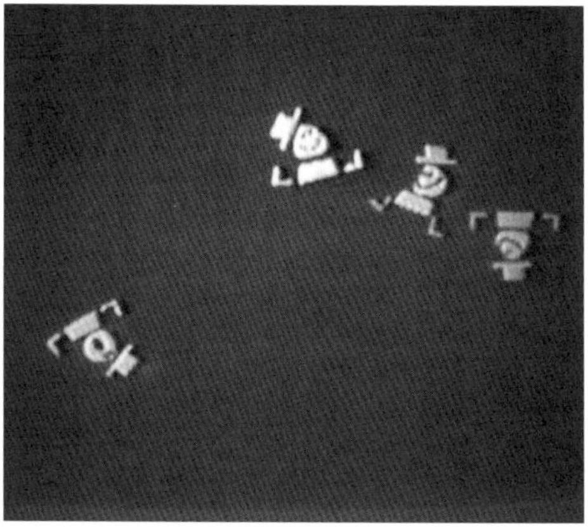

and psychological attachment to an object.

Shin

I have been playing saxophone since
I was twelve years old. Playing this
instrument has left me with fundamental
memories of having a relationship with
an object. Musical instruments are
objects which exist apart from a modern
design context. They are designed purely
for sound. I feel a great satisfaction
playing an instrument like the saxophone.
There is a sensitive, physical contact
with mouth and finger... the sound
vibrates through your whole body.
I can feel the saxophone became a
part of my body...

This is a moment of complete physical

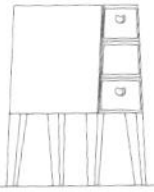

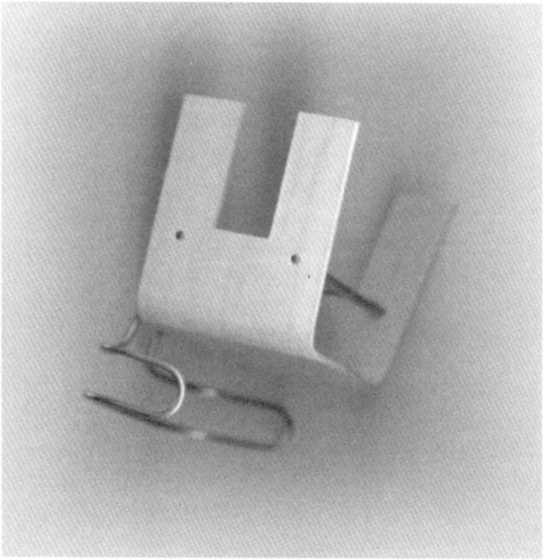

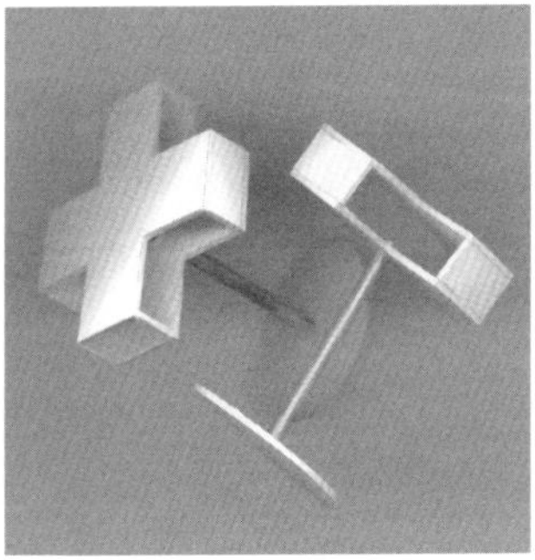

The pleasure and amusement shown by Shin in witnessing someone become aware of the implicit aspects of performance in using one of the Azumis' pepper mills, or our discovery that there is a simpler and easier way to using a piece of furniture we have all been familiar with, but thought little about, is a moment of acute appreciation of the Azumis' work. And going much further than stereotypes of 'Japaneseness' — that 'minimal' quality in their work which may come to mind in first being introduced to, say, *Armchair = Table*, we begin to sense that there is something to be reckoned with here — a finely tuned attention to movement, to playfulness and the exploration of the boundary between abstract shape and 'character' within objects.

The conversion of their 'dual purpose' furniture, in form and character, its emphasis on movement, all impact on the way in which we go about our daily lives. This thoughtful playfulness, when given the opportunity to engage with the Azumis' designs, adds to our understanding of both the object and the environment in which it might be situated — essentially we are given a chance to understand something more of ourselves.

Alexander Payne

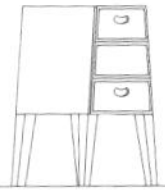

Such observations, and their subsequent interpretation by the Azumis', allow us to interact with their designs, leading us to the discovery of new and more refined approaches to our own behaviour.

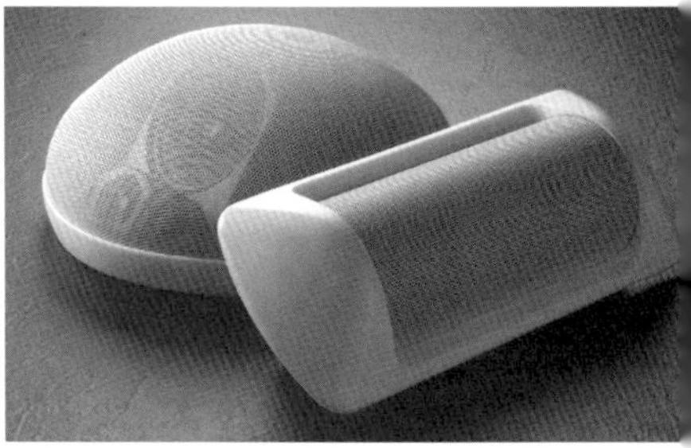

Yet the Azumis' designs, despite their emphases on accessibility, and underlined by a sense of humour and fun, are independent of engagements with and reactions to fashion or the fickle trendiness informing so much of recent design. Rather, not having been hyped by the press, they have been provided with the opportunity of quietly getting on with their own particular vision. As such, their work cannot be located with reference to a specific time or source. Their voice, sometimes misleading in its singularity, is very much of this moment, but also of something more temporal — of a somehow shifting and seamless timelessness.

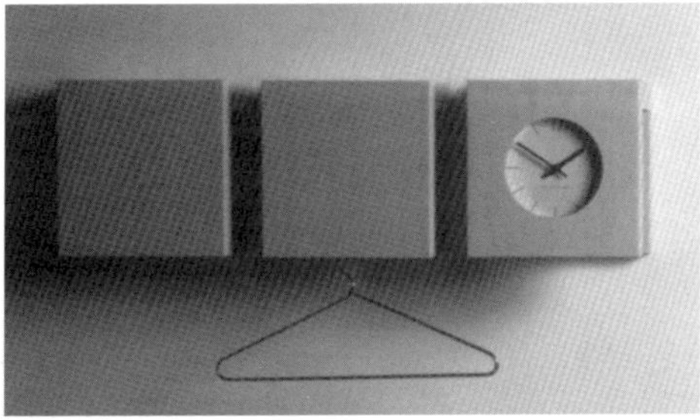

The combination of Shin's observation of daily life and Tomoko's passion for spaces have, through the Azumis' designs — attentive to both function and aesthetics— created a singular and remarkable identity within the world of contemporary design. This unique 'way', an expression of the Azumis' genuine concern for the consumer, can be seen through their meditations on the minutest details of daily life and the transformation of these into objects we can truly participate with. Their 'dual purpose' furniture and the 'performances' subtly encouraged by their elegant tableware, are evidence of the effectiveness of their methods.

"...animation and motion, a mixture of both of our interests—theatricality from Shin, changes from me. We found design to be an overlapping area between us."

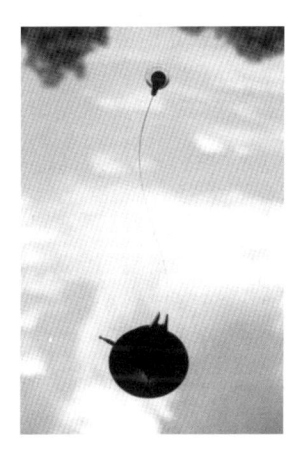

series editor alexander payne

ideas = book

V.2	SERIAL BOOKS
	DESIGN

1-901033-02-3

azumi